LIBRARY MEDIA CENTER
Victor Primary School
Victor, New York 14564

LIBRARY MEDIA CENTER
Victor Primary School
Victor, New York 14564

Eco-Journey

EXPLORING FORESTS

Barbara J. Behm
Veronica Bonar

Gareth Stevens Publishing
MILWAUKEE

For a free color catalog describing Gareth Stevens' list of high-quality books, call 1-800-341-3569 (USA) or 1-800-461-9120 (Canada).

ISBN 0-8368-1064-3

North American edition first published in 1994 by
Gareth Stevens Publishing
1555 North RiverCenter Drive, Suite 201
Milwaukee, WI 53212, USA

This edition © 1994 by Zoë Books Limited. First produced as *Take a Square of Forest* © 1992 by Zoë Books Limited, original text © 1992 by Veronica Bonar. Additional end matter © 1994 by Gareth Stevens, Inc. Published in the USA by arrangement with Zoë Books Limited, Winchester, England.

All rights reserved. No part of this book may be reproduced or used in any form or by any means without permission in writing from Gareth Stevens, Inc.

Photographic acknowledgments
The publishers wish to acknowledge, with thanks, the following photographic sources:
t = top *b* = bottom
Cover: Bruce Coleman Ltd.; Title page: Hans Reinhard/Bruce Coleman Ltd.; pp. 6 Thomas Buchholz/Bruce Coleman Ltd.; 7 John Shaw/Bruce Coleman Ltd.; 8, 9*t* Hans Reinhard/Bruce Coleman Ltd.; 9*b* Jane Burton/Bruce Coleman Ltd.; 10*t* Michael Price/Bruce Coleman Ltd.; 10*b* Rod Williams/Bruce Coleman Ltd.; 11 J. L. G. Grande/Bruce Coleman Ltd.; 12 Gordon Langsbury/Bruce Coleman Ltd.; 13*t* S. Nielson/Bruce Coleman Ltd.; 13*b* John Shaw/NHPA; 14*t* Laurie Campbell/NHPA; 14*b* Michael Leach/NHPA; 15 Stephen Dalton/NHPA; 16 John Shaw/NHPA; 17*t*, 17*b* Dr. F. Sauer/Bruce Coleman Ltd.; 18*t* George McCarthy/Bruce Coleman Ltd.; 18*b* Manfred Danegger/NHPA; 19 E. & P. Bauer/Bruce Coleman Ltd.; 20*t* Hans Reinhard/Bruce Coleman Ltd.; 20*b* Dennis Green/Bruce Coleman Ltd.; 21 J. L. G. Grande/Bruce Coleman Ltd.; 22*t* Andy Purcell/Bruce Coleman Ltd.; 22*b* John Shaw/NHPA; 23 Charlie Ott/Bruce Coleman Ltd.; 24 E. A. James/NHPA; 25*t* Andy Purcell/Bruce Coleman Ltd.; 25*b* Wayne Lankinen/Bruce Coleman Ltd.; 26*t* R. Balharry/NHPA; 26*b* Uwe Walz/Bruce Coleman Ltd.; 27 Henry Ausloos/NHPA.

Printed in the United States of America

1 2 3 4 5 6 7 8 9 99 98 97 96 95 94

Title page:
A tree creeper clings to a tree with its long claws.

Contents

This is the forest 6
In the spring 8
Living among the trees 10
Under the trees 12
Under the forest floor 14
Summer insects 16
Forest hunters 18
Birds of prey 20
Color and camouflage 22
Preparing for winter 24
Winter snows 26
More Books to Read/Videotapes 28
Places to Write/Interesting Facts 29
Glossary 31
Index 32

Words that appear in the glossary are printed in **boldface** type the first time they occur in the text.

This is the forest

Few plants grow on the forest floor because the tall trees block the light.

▼ Only occasionally does light reach the forest floor.

During the long, cold winters, snow lies in deep drifts among the trees. The trees get very little water because the water in the ground freezes.

▲ Many trees in the forest are **conifers**, or trees that have cones. The branches of most conifers slope downward so heavy snow can easily slide off the trees.

In the spring

In the spring, **pollen** from male cones of one conifer tree **fertilizes** the seeds inside female cones of another conifer tree.

▶ In the spring, the Scotch pine grows woody cones in clusters at the end of its branches.

◀ Crossbills, which live high up in the trees, eat pine seeds.

▼ Birch catkins hang down from the branches of silver birch trees. Pollen is carried from one catkin to another by the wind.

Conifer leaves contain a sticky, strong-smelling substance called resin. It is poisonous, so the leaves are safe from being eaten by most birds and insects.

Living among the trees

As the weather gets warmer in the forest, insects hatch into **larvae**.

▲ The caterpillar larvae of the pine sawfly gather in groups of up to a thousand. They eat the young pine needles and damage the trees.

▶ In spring, a chipmunk comes out of its underground nest where it has spent the winter.

When the snow has melted, small **mammals**, such as chipmunks, search for insects, seeds, and nuts to eat. Birds feed insect larvae to their young.

▲ Jays make a cup-shaped nest of twigs to care for their young.

Under the trees

During the winter, deer eat bark from trees. In warmer weather, they feed on mosses and **lichens** off the cool, damp forest floor.

▼ In spring, male deer grow antlers covered in soft, furry skin.

◀ Mushrooms grow well in the dead leaves and pine needles on the damp forest floor.

▼ Young leaves on the trees and moss on the rocks beneath the trees provide food for forest animals.

Ferns, mushrooms, and other **fungi** also grow in the damp soil.

Under the forest floor

Wood ants leave their underground nest with its many tunnels and chambers to collect insects and seeds. They drag the food back to the nest to feed the queen and larvae.

▲ Wood ants build a pile of pine needles to hide their nest.

▶ A shrew searches for insects and worms. Shrews cannot live for more than a few hours without food.

Small animals, such as voles and shrews, build tunnels under the snow, where they spend winter.

▲ Voles store seeds, roots, and leaves for the winter.

Summer insects

Wasps come out to feed in the warm summer sun. The ichneumon fly, with its thin, pointed tail, also appears.

▶ An ichneumon fly looks like a wasp.

◀ The sawfly feeds on the sweet nectar of flowers.

▼ The pine weevil uses its long snout to bore through wood. These beetles chew the bark of young trees, which damages the trees.

Sawflies lay their eggs once a year under the bark of trees. The eggs hatch in the spring.

Forest hunters

Small animals are hunted by larger animals called **predators**. Weasels and pine martens are predators. They hunt mice, voles, and rabbits.

▲ Weasels mark their hunting grounds with their scent. This warns away other weasels.

▶ Pine martens mate in the summer. Between two and four young are born during the following spring.

Larger predators, such as wolverines and bobcats, hunt rabbits, mice, birds, and squirrels. Their thick fur keeps them warm in winter.

▲ A bobcat has long legs and large paws. In warm weather, it hunts squirrels and birds. In winter, a bobcat will catch whatever food it can find.

▼ A merlin, which is a falcon, sits motionless on a branch. It watches for any small movement on the ground below.

Birds of prey

Hawks, owls, eagles, falcons, and goshawks are called **birds of prey**. They have keen eyesight and catch their prey with powerful talons, or claws.

▶ A peregrine falcon can fly at speeds of up to 150 miles (240 kilometers) per hour toward its prey. It is the fastest creature in the world.

Large birds of prey hunt in open spaces. Their wide wingspan makes it difficult for them to fly among the trees.

▲ Golden eagles make their nests at the tops of the tallest trees or rocks. Both parents search for food for the young.

Color and camouflage

▶ When this moth rests, it closes its wings over its back to provide camouflage.

▼ In summer, the brown feathers of the ptarmigan hide it on the ground. In the winter, its feathers turn white to hide it against the snow.

Animals that live on the forest floor are usually brown. Their color blends in with the forest floor, **camouflaging** them from their predators.

In autumn, some of the forest birds and other animals become white in color so that it is harder for predators to see them in the snow. Some predators, however, also turn white.

▼ The ermine preys on mice, voles, and shrews. Its fur turns from brown to white in winter.

Preparing for winter

As winter draws near, certain birds fly south to warmer lands. The rest of the birds and other animals begin to store food to last the winter.

▶ Many birds fly south to warmer lands in winter. There, they may find berries to eat.

◀ The snowy owl is well prepared to survive in the cold.

The snowy owl can survive cold winters. Its keen hearing allows it to hunt small mammals moving around in their tunnels beneath the snow. It swoops down silently on its prey.

▼ Black bears may **hibernate** as much as six or seven months each year. As they sleep, their heartbeat slows and their body temperature falls a few degrees.

Winter snows

Snowshoe hares survive the winter by nibbling birch bark and small trees. Foxes hunt hares and other small mammals that live in burrows beneath the snow.

▲ Snowshoe hares have large, hairy feet and wide back legs that prevent the hare from sinking into the snow.

▶ Foxes can smell a mouse or a vole under deep snow.

Wolves hunt large prey, such as deer. To keep warm in winter, wolves rest in holes in the snow.

▲ Wolves hunt deer over long distances through the forest.

More Books to Read

Calling for Our Forests. Carol Greene (Enslow)

First Look in the Forest. Daphne Butler (Gareth Stevens)

How the Forest Grew. William Jaspersohn (Morrow)

Life in the Forest. Eileen Curran (Troll)

Our Changing World: The Forest. David Bellamy (Crown)

Trees. Sharon Gordon (Troll)

Videotapes

Call or visit your local library to see if these videotapes are available for your viewing.

The Enchanted Forest (An old hermit teaches a young boy to love the forest and its creatures.)

The Wonderful World of Disney: The Ranger of Brownstone (with J. Audubon Woodlore and D. Duck, followed by live-action footage of birds).

National Geographic Series – The Grizzlies.

Places to Write

For information about Audubon nature centers in your area, contact:
National Audubon Society
700 Broadway
New York, NY 10003

For more information about forests and wildlife, contact:

U.S. Department of Agriculture
    Forest Service
Public Affairs Office/
    Publications
Auditors Building 2 Central
201 Fourteenth Street, S.W.
Washington, D.C. 20250

Internal Ministry of
    the Environment
Public Information Center
First Floor
135 St. Clair Avenue, West
Toronto, Ontario M4V 1P5

Interesting Facts

1. The seeds of pine trees and other conifers have papery wings, so the wind can blow them away from the tree to a new area where they will sprout.

2. Birds do not eat only adult insects. They feed on larvae, or insect eggs. In this way, they keep insects from taking over the forest.

3. Some blue jays steal eggs from the nests of other birds.

4. When the nest of wood ants is disturbed, the ants inside squirt an acid into the air to warn nearby ants of the danger.

5. When there is plenty of food, voles may produce up to eight young every three weeks. When there is less food, they have fewer babies and less often.

6. Weasels usually hunt alone at night. They use their keen sense of smell and sharp eyesight to find their prey.

7. Weasels are small enough to follow mice and voles right into their burrows.

8. In the United States and western Canada, timber fires may last for weeks.

9. In the United States and Canada, helicopters and airplanes dump water and chemicals on forest fires to put them out.

10. Ecologists think that natural fires caused by lightning may often help forests because they clean out brush and unhealthy plants.

Glossary

birds of prey: meat-eating birds, such as hawks and falcons.

camouflage: the color and shape of the fur, feathers, or skin of an animal that make it difficult to see against its background.

conifers: trees that bear seeds in cones.

fertilize: to join male cells and female cells so new plants grow.

fungi: plants that cannot make their own food.

hibernate: to pass the winter in a resting state.

larvae: the second stage in the life of insects, after the eggs have hatched.

lichens: two plants that live together, a fungus and an alga, and help each other survive.

mammals: animals that carry their young inside their bodies before birth.

pollen: tiny spores produced by plants as a fine dust.

predators: animals that eat other animals for food.

Index

bark 12, 17, 26
beetles 17
birch 9, 26
birds 18-21, 23, 24
black bear 25
bobcat 19

camouflage 22
caterpillars 10
chipmunk 10, 11
cones 7, 8
conifers 7, 9
crossbills 9

deer 12, 19, 27

eagles 20, 21
eggs 17

falcons 20
ferns 13
foxes 26
fungi 13

golden eagle 21

goshawks 20

hares 26
hawks 20

ichneumon fly 16
insects 9-11, 14, 16

larvae 10, 11, 14
leaves 9, 13, 15
lichen 12

mammals 11, 18, 25, 26
mice 18, 19, 23, 26
moss 12, 13
moths 22
mushrooms 13

nests 10, 11, 14, 21

owls 20

pollen 8, 9
prey 20, 21, 23, 27

ptarmigan 22

rabbits 18, 19
resin 9
roots 15

sawflies 17
seeds 8, 9, 11, 14, 15
shrews 14, 15, 23
snow 7, 11, 15, 22, 25, 26
snowshoe hare 26
snowy owl 25
squirrels 19

voles 15, 18, 23, 26

water 7
weasels 18
wind 9
wolverines 19
wolves 19, 27
wood ants 14
worms 14